# Dream Big

# Dream Big

Don Ross

**To order additional copies of this book, contact:**
Xlibris Corporation
1-888-795-4274
www.Xlibris.com
Orders@Xlibris.com
83457

# Bibliography

I am listing some functional accomplishments so far:

## 1) MOST LIKELY THE FIRST F-M RADIO STATION ON UNIVERSITY CAMPUS

I went over to see a Professor about a FM Radio Station on campus. He told me all its merits. I asked him how long was it going to take to get the Radio station on campus? His response was next year or never. I went back to the dorm to study. An idea came to me. A talk-a-thon around the clock might get the Radio Station earlier. It's an interesting story. The Radio Station was on campus the NEXT year and we had an exchange dinner with the girl's dorm.

## 2) VOLUNTARY MILITARY SERVICE

I had an idea while in ROTC on the Drill Team. Let's have Voluntary ROTC and the attitude of the Troops would change. We went early to drill and presented the idea to the Sergeant in charge who was a West Point Graduate for six months. The University was a Land Grant University which in the US Constitution requires compulsory ROTC. As we were finishing our required second year. They told us that Congress granted two years of Voluntary ROTC. They said the first year enrollment would go down 200 men and the second year enrollment would go up 200 men and in your face. But, the first year enrollment went up 200 men and the next year enrollment went up 200 men and the attitude changed and so did productivity. Congress was impressed and made Voluntary service Military wide.

## 3) REAL TIME COMPUTER OF TODAY

While working at Control Data Corp. I was asked by General Ford for the CEO if one of his possible future protect would work as follows. We were told in a training class that we would be asked how to make the large Mainframe Computer better for we had six months for our Clearance to go thru. Another interesting story. The General asked me where is Industry going. My response was Real Time. I went to a TRW site in California. Two Phd's were brought in one on my side of the answer and the other on existing batch computers. Real Time won out and you have the Computers World Wide that you have today. It was an interesting story. The academic world and all Military were watching and the CEO. My mentor General Mc Arthur's Chief Engineer told me this is what the Corporation would do. Another interesting story.

## 4) MULTIPLE 60 BIT WITH DOUBLE PRECISION 120 BIT PROCESSORS IN A SINGLE COMPUTER

The CEO had another project he was thinking about. Another interesting story. The question was asked me by CDC's Chief Software Person on the TRW site if a Computer with 256 Central Processors would work. My answer was Yes. The next weekend a Phd. was brought in to build a Large Mainframe Computer with 16 Central Processors and 16 I/O Channels. Six months later I was shown the Computer. The Computer grew to 1024 Central Processors in a letter to IBM. Then over 70,000 CPU's and the Human Genone was mapped. Then it grew to over 140,000 CPU's and Gene Therapy was possible and other great things. I was offered a VP position. Later I was told that they wanted me to get experience and come back and they would pay me what I was worth.

## 5) TWIN ROVERS TO MARS

When the Shuttle blew up, I wanted Daniel Golden the Head of NASA to fix it. Another interesting story. They fixed it and a line of communication was open. They were going to send either a orbiter or a rover to mars. In a letter I asked for both. After four days he announced Twin Rovers. After a letter about possible identical Triplets to the Head of NASA.

## 6) INVENTORY CHIP

While working At Walley World (Walmart) to bring in some money while working with our special needs daughter for 36 years. An idea came to me. There was a 5 cent chip which could be imbedded in the products at Walley World. The chip would communicate to the Computer System and inventory could be done in less than a hour. Scientific American wrote an article in their magazine. The chip could be scanned at Check Out the buggie and the person. A printout would charge the result of the scan. The Lawyers got involved and said this was an invasion of privacy and stealing was ok. The project was unfortunately dropped.

### 7) New Buzzie Rocket and New Lift System

While working with our Special Needs Daughter, I had time to test out many projects. The New Buzzie Rocket and New Lift System were developed independently thru our Corporations. The New Buzzie Rocket is better than any Rocket today. VonBrun's first Rocket went up and turned down and blew up on the ground. Another interesting story. Our Corporations tested the New Buzzie Rocket and New Lift System on a micro scale with fantastically great results. We need funding! The New Buzzie Rocket eliminates the costly and slow turn around times, guidance system, etc. The New

Lift System would reduce the Weight of the Space Shuttle to ONE POUND.

## 8) Three Questions

If the Space Shuttle was reduced to ONE POUND and stays there during launch and orbital time by a computer by my New Lift System, BETTER THAN the Lift System developed by 60 Polish Scientists at the end of World War II. What would be the VELOCITY OF THE SPACE SHUTTLE THEN?

How long would it take at this VELOCITY TO TRAVEL THE DISTANCE TO THE MOON?

Would this bring about ARTHUR C. CLARK's 2010?

Feb. 7th, 2010

I was in High School which was ranked 5th in the USA. An interesting Heavenly thought was given me to my subconscious. Along with John Hatfield of the Hatfield's and McCoy's to do an experiment. It was 8:00 p.m. the night before Christmas coincidently. All in 1958. What a beautiful Christmas Present! An unexpected impossible I thought 5.6 percent reduction in total weight of two objects!!!!! Later, after studying Einstein's theory's of Relativity and the result of vaporizing the Protron Core which yielded a Positron and Electron in the text books. It all made sense! A Scientist friend did my experiment and got a result of 7 percent reduction in total weight! Without my experiment which was not published as of now. The Protron Model is incorrect! The Pioneers of Atomic Theory made incorrect assumption that there was not an Electron on the Core of the Protron! The assumption has led to a Physics mess of today!!! The LHC has to break again to prove that a beautiful gift of CLEAN ATOMIC FUSION IS POSSIBLE!!!!!!!! It would be wise not to be in the LHC tunnel when it reaches a certain energy level! It's just Physics!!!

State of Colorado, County of

_Arapahoe_____, I,_____Seth   Thompson_____,a

Notary Public in and for said state, do

certify that on __6/22/10__ , I carefully

compared with the original facsimile of

A Book and the facimile I now hold in

my possession. They are complete, full

and exact facimiles of the document they

purport to reproduce.

NOTARY PUBLIC
SETH
THOMPSON
STATE OF COLORADO

My Commission Expires 03-03-2012

_____3/3/2012_____

# Chapter 1

Do two magnets lose weight when they are joined together?

It is good to start at the beginning.

The beginning of this book was in the year 1958. It was a great year!

Do two magnets lose weight when they are attached?

I was working in the Aerospace Industry. While watching TV to unwind after my evening—night shift when a God given idea came to me. Does the weight of two magnets equal the total of the calculated total weight when the two magnets attached? I decided to sleep on it and go to my father's assay office the next evening which was the night before Christmas. The next day I ran into John Hatfield of the Hatfield's and McCloy's and asked if he would like to see an experiment at my father's assay office? John said yes. I drove to my father's office.

It was good to have a person with you when you make a big breakthrough in the laboratory to see the results. John wore a three piece suit with work boots which was an interesting site. I weighed each magnet and the cardboard separators separately very accurately on the analytical balance. {2} Diagram for Analytical Balance is in the Index. The moment of truth was about to happen! Would the two magnets weigh the calculated weight or not? This was a great moment of a Scientific Breakthrough. This was equal to the famous E=m c squared equation? Their weight and the separators were a whopping 5.6% less than the calculated weight—WOW!!! What a Christmas Present!!! Then the experiment proceeded with the more cardboard separators added. The weight was the calculated weight. More cardboard separators were added and the weight was 5.6% greater! I did not understand this at the time and put it in the file to know at a later time. I closed the Office and drove John to his apartment and went home to get some sleep for the next day was Christmas!!! The idea came from God!!! Some doubt it, but I have talked to many original discoverers and they all agree that their discoveries came from God!!! Have a great day!!!

# Chapter 2

Would many combinations yield the same results.

I decided that I would go back to my father's office with five magnets, the results of which would be calculated later. {3} Five magnets diagram in Index. The magnets were weighed in two's, three's, four's and five's at a time. Two were weighted in attraction and two in repulsion with duct tape to hold them together!!! The 5.6% less held for both attraction and repulsion as did the calculated other combinations. The different sets of two all were 5.6 less also. All sets of two with more cardboard separators agreed with the actual calculated weight when separated and when separated further all agreed with the 5.6% greater.

# Chapter 3

Does an iron bar become heavier when magnetized?

The experiment was set up with a broom stick without the broom head attached.

{4} The broom stick balance is in the Index. The broom stick was suspended from the ceiling with a rope on a hook in the ceiling with a pointer to read the balance on an Architect ruler given to me by another Engineer. A light was shown though the pointer to the ruler to get the reading. On other end was the bucket for weights. The iron bar was weighed. Next, the iron bar was magnetized and weighed again. It was 5.6% greater.

Now, I knew the magnet fields were interacting with the two magnets and explained the observed three weights observed. Another idea came to me. Do magnetizable fences, poles, etc. in Littleton and elsewhere get magnetized in the Northern Hemisphere? I set out around Littleton, Denver and in the mountains to see if this was the case. I checked the polarity of the objects with a magnet on a string. {5} Magnet on a string. Diagram is in the Index. The above objects were magnetized the same polarity to my amazement! No doubt, objects in the Southern Hemisphere would be the opposite magnetic pole.

# Chapter 4

Uncovering the mystery of the Protron.

The decrease of 5.6% in weight is the due to the interaction of the magnetic fields. This was the first clue that the electron was hidden under a positron cloud on the heavy core of the protron. In the beginning the Scientist's calculations would not show the electron, but when they vaporized the heavy core with an 1.2 mev beam, an electron and positron would come off and go into opposite orbits in a magnetic field and the positron would interact with its counter part and give off energy according to $E=m c$ squared. Then, the Positron appeared again only without its positive charge or its charge was neutral! The electron field would interact with the positron field and would decrease the protron's weight and mathematically show that the electron was there on the solid core of the protron. There is a picture of this with the earth's negative charge on its surface and its positive charge on the outer surface of the two Vanallen belts. The earth would be the solid core. The protron is made up of 1) positron, 2) an electron, 3) a solid core. {6} The Protron Diagram is in the Index.

# Chapter 5

Some things that happened along the way and a short study of Ezekiel" vision and how it shows a System which has yet to be build today, but was seen nearly 2500 years ago!

A BS Physical Science in Math and Physics was completed, a BSEE in Electrical Engineering was Completed, and 1/3 of a MSEE in Electrical Engineering and Computer Science was completed At CSU, DU, and CU. I got Married and moved to California from Minneapolis, Minn. where I completed my Mainframe Computer course. Next, I refer to Ezekiel Chapter 1 Verse 4. God save me the idea to read and study in detail Ezekiel's vision from an Engineering and Scientific view along with experiments to test what was observed. This is the reason for this first book. If young people are to build and fly the System that Ezekiel saw nearly 2500 years earlier, and it is a great blueprint!!! Now, I go to what happened in this study and the following experiments! It was my approach to look at Ezekiel's vision in the King James version of the Bible for it is the closest version to the original text. The first step was to build an index of every word in his vision and in which verse the word was found. Next, there were two things that were not clear until I went back to very old English. Eyes refer to orifices or pupils which made the System very contemporary! {7} Eye Diagram is in the Index. Orifices are used in most modern design of Air Conditioning, Electrical, Mechanical and all Engineering including Digital, Etc., Architectural design of Cruise Ships, Buildings, Homes, Cars, Trucks, and Bioengineering to help in all medical breakthroughs!!! I had asked engineers what up and over was meant to no avail. {8} Up and Over Diagram is in the Index.

When I went back to my fluids course and book it become very clear! It showed how my New Buzzie Rocket is very much better than any Rocket in History!!! Now, I will go through the main ideas and show what gems I found in Ezekiel's vision. I did 40 years of research and used Ezekiel's vision as a guide along with the New Lift System of World War II, which I improved! {9} The New Lift System Diagram is in the Index. This is what I call the Buzzie System and it is marvelous!!! {10} My Buzzie Story is in the Index. {11} The Military feedback Diagram is in the Index. Military feedback and my calculations show that it's probable that the distance to the Moon can be covered in less than a half an hour and the distance to Mars can be covered in a week!!! Would this bring Author Clark's 2010 about??? Now to start your fantastic contemporary study (fasten your seat belts!!!) and Enjoy!!! After this study I hope that you come to the same conclusion that I did that Ezekiel in his vision saw a System that we have not built it yet! His description is a good blueprint to follow so that you do not have to go off on any Rabbit Trails!!!

We will start with Chapter 1, verse 4 and go very quickly!!! First we see the Reverse Diode Effect is most probable in improving the New Buzzie Rocket Engine's velocity and thrust and decreasing the travel time as was stated earlier!!! {12} The configuration Diagram is in the Index. of the four creatures yield 75% greater lift than any plane today as seen in my experiments. The form of a man also increases lift. The wings enable the System to fly subsonic and supersonic. Next, we see international language of sculptures to identify the flight that you booked. There are two Systems one for Earth flights and the other for Space business and tours to the Moon and Mars, etc. It will be very exciting (Do you have your bags and suit cases packed?)!!! Their feet provide at least three functions support, Lightning arrest to ground and less drag in flight.

# Chapter 6

Many interesting things in the 40 years of study!!!

Next, I will present some of the many interesting things that happened during the 40 years of this interesting research. One of my Heroes from fifth grade was Walt Disney who said: "if you can believe it you can do it". Walt had faith!!! Think of arriving on the moon and derocketing with protective shields to have fun, sports, entertainment, Universities, fantastic telescopes, etc. for your stay!!!{21} The Bubble Diagram is in the Index. (It is awesome how fast you could fly to the Moon!) LET'S GO FOR IT!!! The Buzzie International Project which has to be and would probably be: many stories tall. How tall is dreadful? {13} Creature Diagram is in the Index.

It would carry very large cargos, Disappearing disks are a fantastic hydrogen generator!

{14} Disappearing Disks Diagram is in the Index.

Also, I observed disappearing magnets which are too a fantastic hydrogen generator too! {15} Disappearing magnet Diagram is in the Index. 4) While working in 1969 as a subcontractor at TRW for CDC Corporation the CDC software head there with CDC asked the question which was from CDC's CEO. I said yes it would work!!! The wonderful computer with 16 CPU's and 16 I/O channels would with double precision 120 bits each CPU started and grew to 288 Multi-Processors at Bell Labs to 1012 Multiprocessors to 32,072 Multiprocessors to 32,072 Multiprocessors to 72,240 (which meant the human genome could be mapped!) to over 131,000 Multiprocessors (Which meant we could do Gene Therapy! Now maybe the Super Grid will have over 131,000 Multiprocessors in each node!!! The CEO of CDC gave me the prints and diagrams to build the first personal computer when I left CDC ten years before the Apple computer was built. Then, he had a pet project in mind. He offered Customers the option of use of a Personal Computer, a Minicomputer, or a Large Mainframe Computer and a charge only for the time you used without having to buy any of the three computers! They were supplied over the telephone lines. A substation burned down and it cost him one million dollars in contracts! He then went to satellite to supply the service to his customers. This was probably the first type of Internet! What is your Dream? 5) Next, a calculated 40 to 6800 or more tons of lift per square foot of surface area. Would be more than enough to bring about the Dream with a fraction observed to date. A technique to come closer to the top figure would work! 7) Red Hot-useful available 28.9898 Million Horsepower or about 21.62 Billion watts as demonstrated in 1899. This was tested in my home lab at 7 million 200 watts!!! The Colorado Springs, Colorado, test put the town in darkness which was fixed the next day! Nikola Tesla and his crew installed an armature which happened to be a spare in the back yard of the power company from which he invented and developed the AC Power System! 8) I observed Iron and Aluminum sublimation into hydrogen gas and I ignited it with a 7200 degree F temperature result and the test broke the test tube in half! {16} Test tube in Half Diagram is in the Index. I observed Aluminum going into hydrogen and back to a solid with water in the a damped 60 cps to approximately 50 cps which was very moving! Aluminum to Hydrogen and back I ignited the Hydrogen with an Electric Arc four times!!! {17} The Aluminum to Hydrogen and back Diagram is in the Index. This explains where the missing mass of the Universe is!!! It took 28 years to solve the above and solve the mystery of the disappearing disks. 9) I built a magnetic monopole and observed it oscillating in the Earth's magnetic field for days!!! {18} The Monopole Diagram is in the Index. I observed the anti-matter curve on my oscilloscope in our kitchen experiment!! {19} Anti-matter Curve is in the Index. 10)

It is interesting that 57 % of the YOUNG PEOPLE (and that does not take into account the Young at Heart!) would like to go to the Moon and Mars!!! Imagine all the neat and fun things that you could do on the Moon! {20) 120 lbs to 20 Lbs. diagram is in the Index. If you weighed 120 pounds, then you would only weigh 20 pounds on the Moon! Imagine how far you could throw a baseball on the Moon within a hemisphere a quarter of facsimile in diameter and a floor 6 feet thick with two magnetic fields to simulate the Vanallen belts on Earth to protect you from the lethal radiation on the Moon's surface. There would be a cloud or dark surface on the top of the hemisphere bubble to provide shade during the day from the Sun's burning rays. {21} The Bubble Diagram is found in the Index. {22} Fish Diagram is in the Index.

11) Produce, animals, fish and most plants would grow up to 6 times larger. I watched Space Crystals used by the astronauts grow larger in my special system used to

simulate one half gravity. Also, I observed plants grow bigger and in a simulated one and a half gravity in my special system the crystals and plants grew smaller. You could allow a cow to grow to the size of an Elephant. The things above could be stored at minus 240 degrees F and used on the Moon or shipped back to Earth!!! The same thing could be said for Mars only things would grow only three times larger, but you could do many fun things on Mars!!! The hope again is this with all of its return will come about!!!

P.S. A special thanks to Judy my wife for correcting my spelling and proofreading the book!!!

Don

A relative of Betsy Ross

And a descendant of Ian Flemming who discovered penicillin!!!

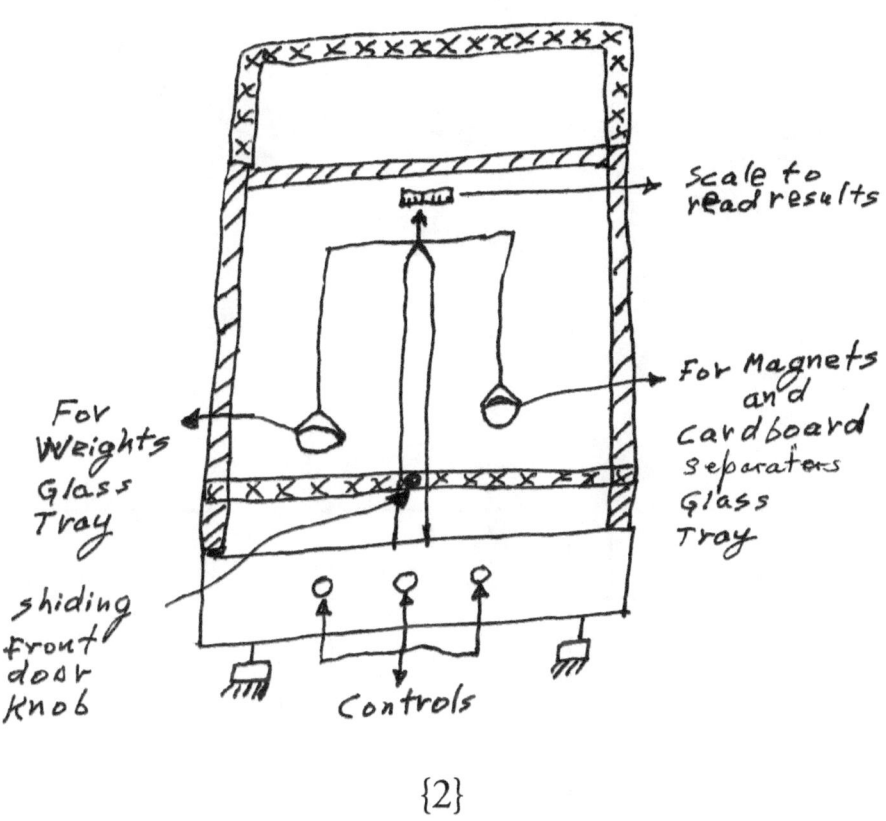

{2}

The Analytical Balance is enclosed in glass & wood to prevent wind currents form throwing off the accurate readings.

Balance Diagram

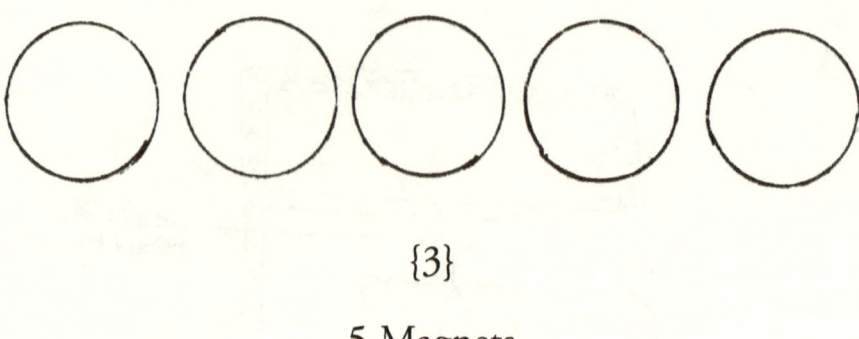

{3}

5 Magnets

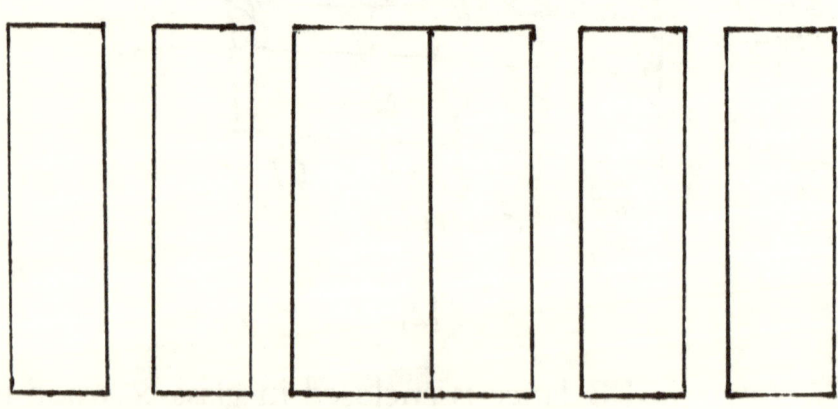

Many Cardboard Sepa

5 Magnets Diagram

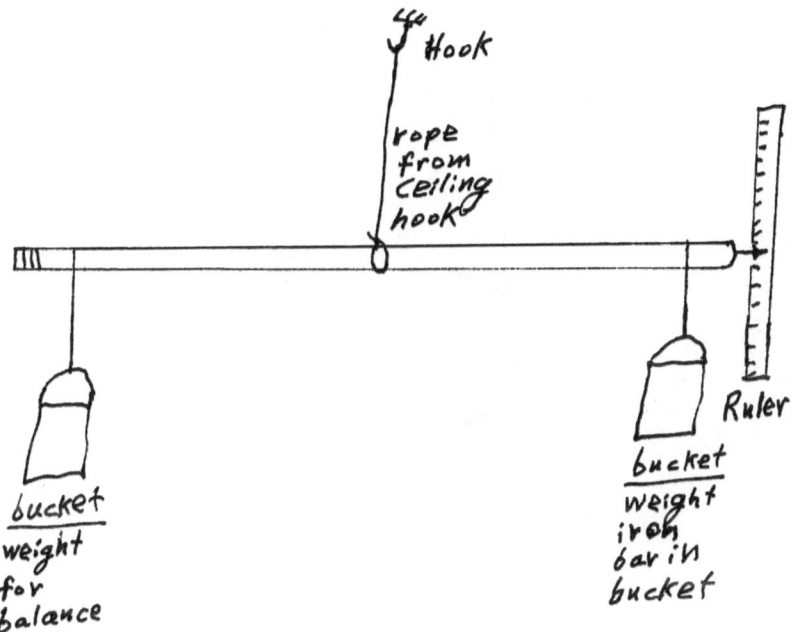

{4}

Broom Balance Diagram

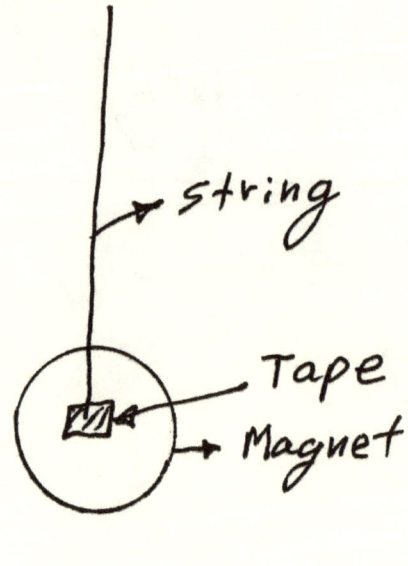

{5}

String and Magnet Diagram

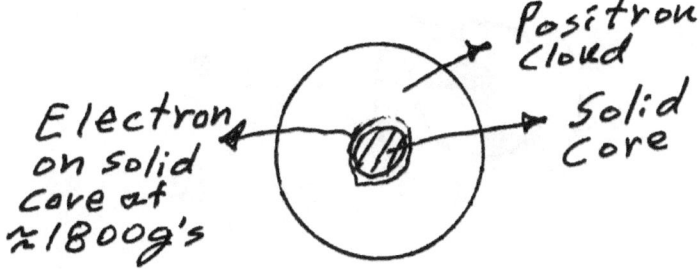

{6}

Protron Diagram

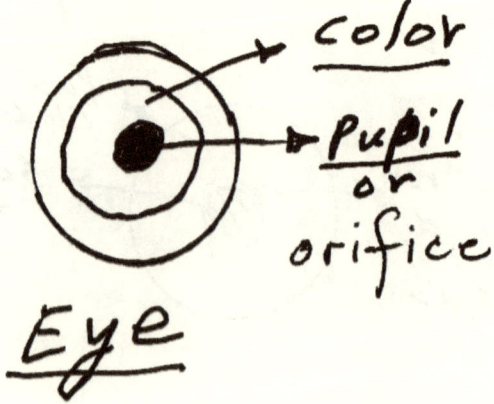

{7}

Eye Diagram

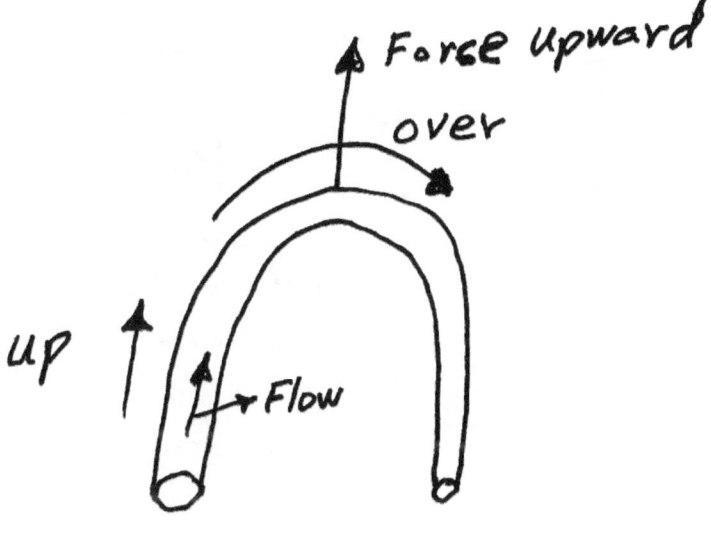

{8}

Up & Over Diagram

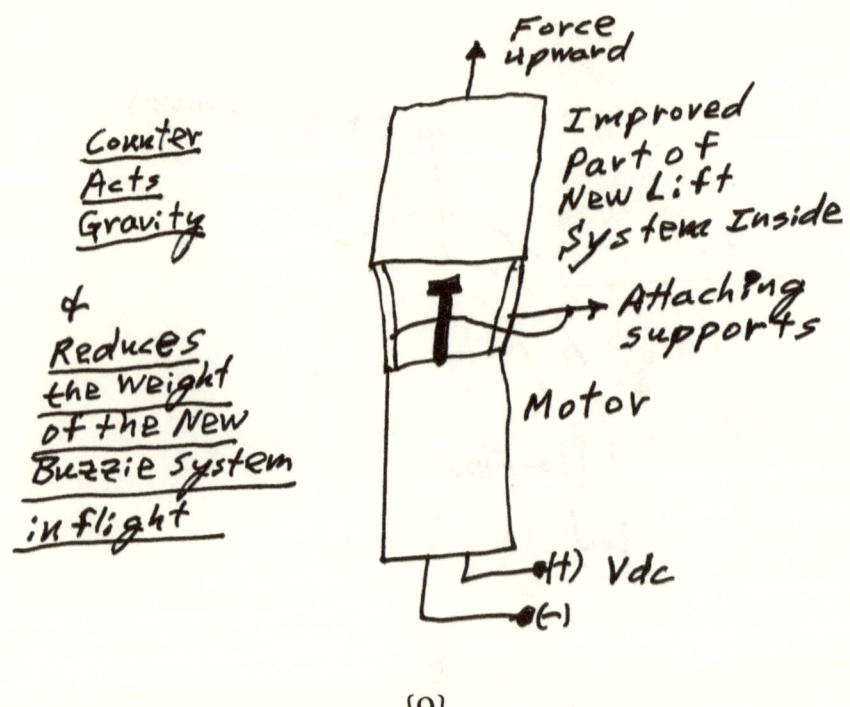

{9}

New Lift System Diagram

## {10}

I sent a rotating model for a store to Disney along with my Buzzey Newsletter to Michael Eisner, former head to Disney in a pyramid cardboard box by UPS. It got through and it was sent back and looked like it was bounced off of a wall. After this, Buzz Lightyear toys and movie came out! I was happy for the kids!

My Buzzie Story

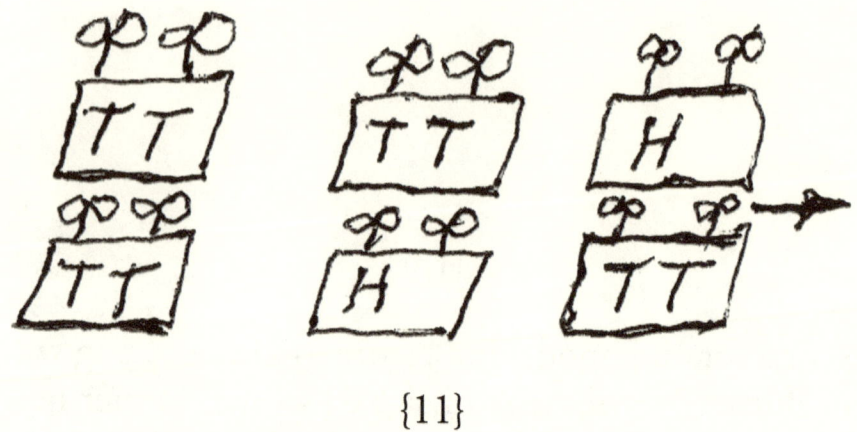

{11}

H equals a Huey Helicopter.

TT equals a Troup Transport helicopters.

Six above Helicopters flew over our Condo at about 30 Miles per Hour. The formation with the position of the Huey and the two TT's said that the Resultant Force upward on the top and bottom of my New Buzzie Rocket were tested equal!!!

Military Feedback Diagram

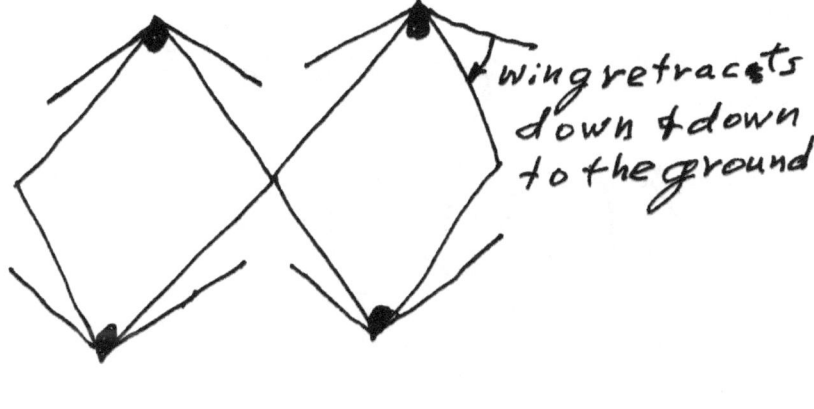

wing retracts down & down to the ground

{12}

Configuration Diagram

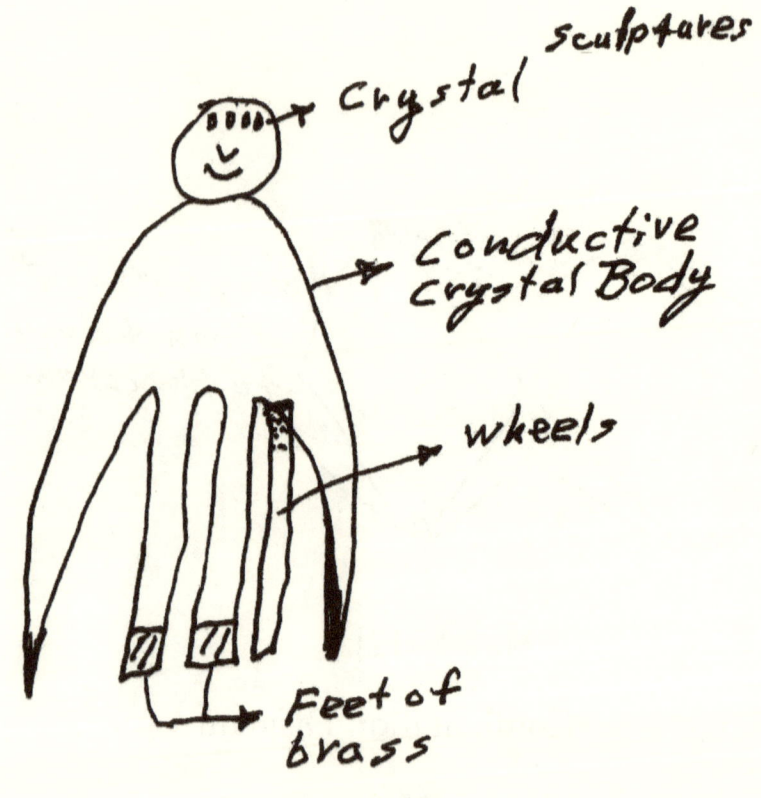

Crystal sculptures

Conductive
Crystal Body

wheels

Feet of
brass

{13}

Creature Diagram

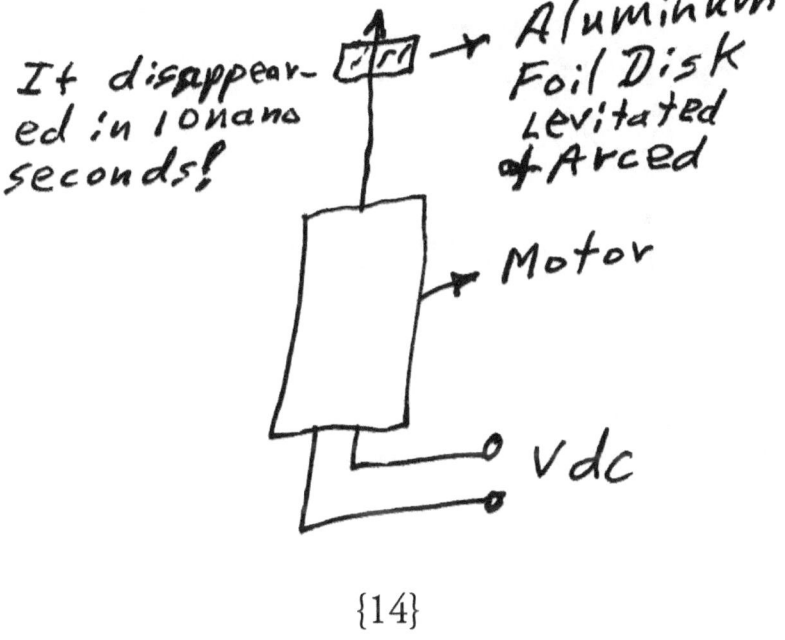

It disappear-
ed in 10 nano
seconds!

Aluminum
Foil Disk
Levitated
~~&~~ Arced

Motor

Vdc

{14}

Disappearing Disk Diagram

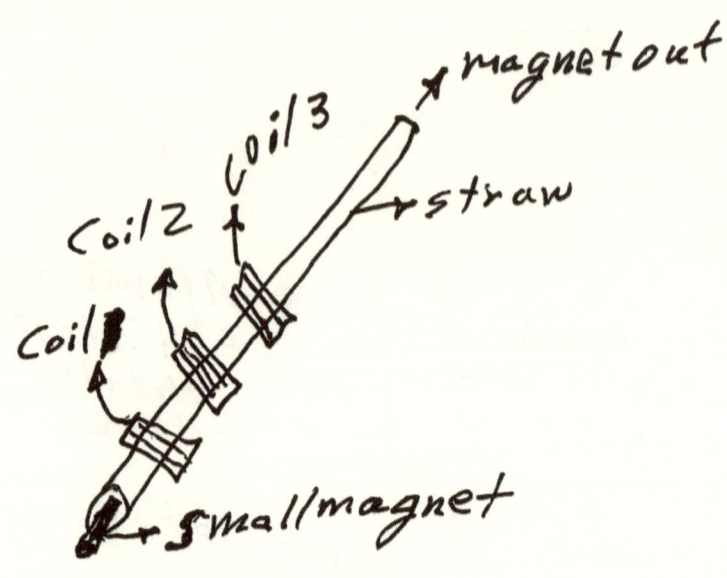

{15}

1) The magnet with coil 1 was launched 3 feet?
2) The coil 1 & coil 2 launched the magnet 10 feet?
3) The magnet did not come out with coil 1, coil 2 and coil 3.
4) I found out later that coil 3 was in backwards and stopped the magnet!!!
5) I checked the straw, when the first magnet disappeared. I launched a second magnet and the first & second magnets were stuck together in the straw!

Magnet Diagram

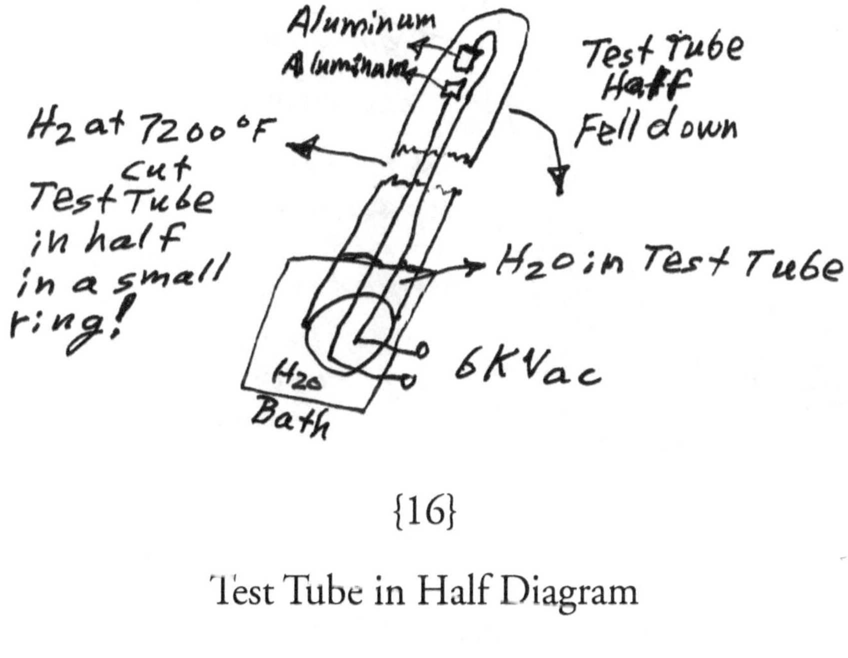

Aluminum
Aluminum

Test Tube
Half
Fell down

$H_2$ at 7200 °F

cut
Test Tube
in half
in a small
ring!

$H_2O$ in Test Tube

6 KVac

$H_2O$
Bath

{16}

Test Tube in Half Diagram

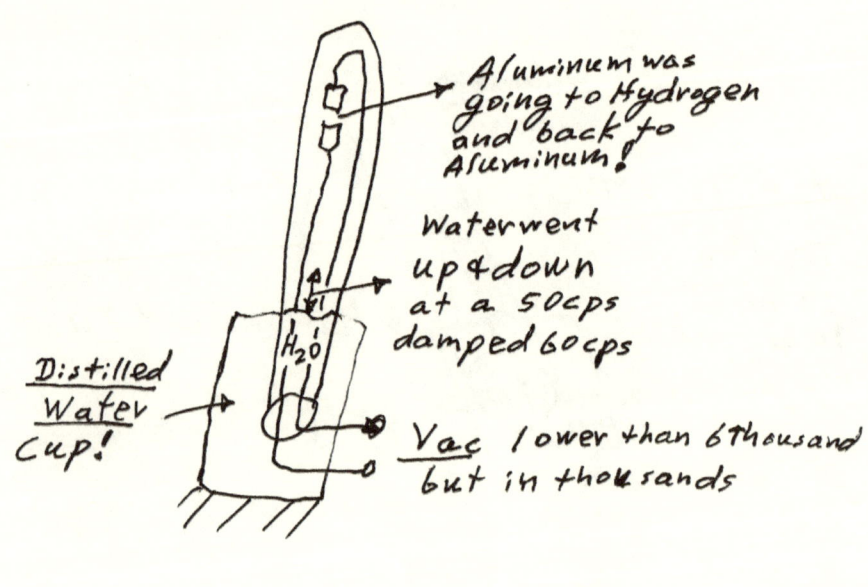

{17}

AL to Hydrogen and back Diagram

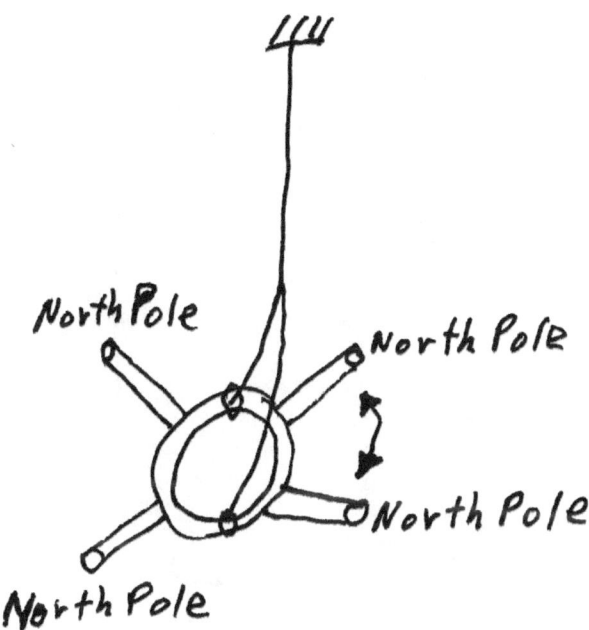

{18}

Monopole Diagram

o scope

{19}

Antimatter Diagram

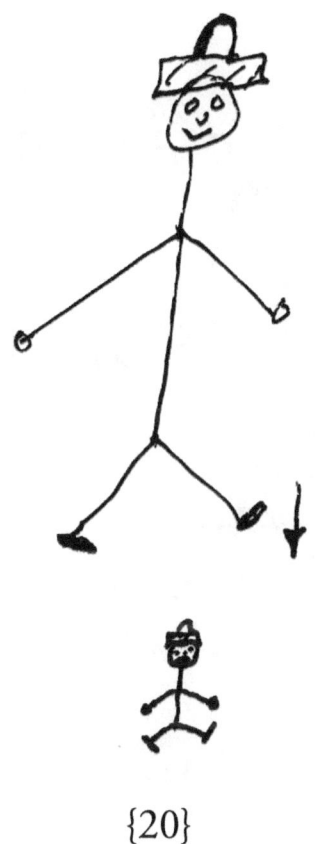

{20}

120 # to 20# Diagram

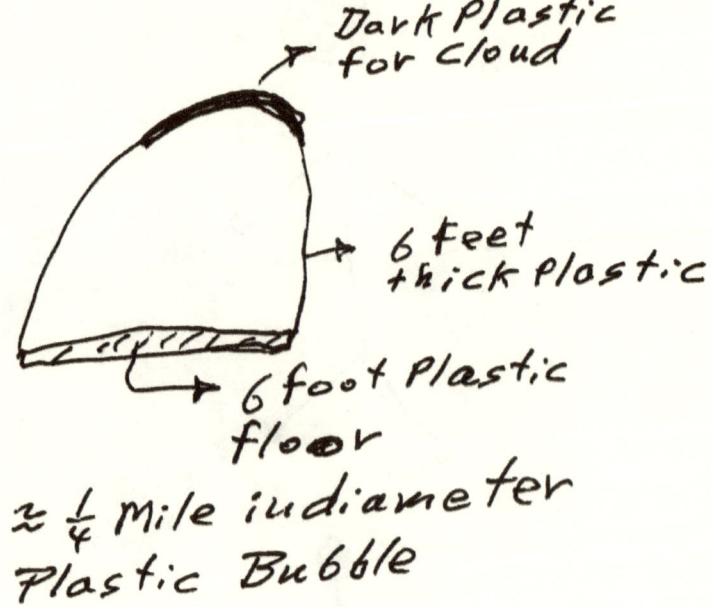

{21}

Bubble Diagram

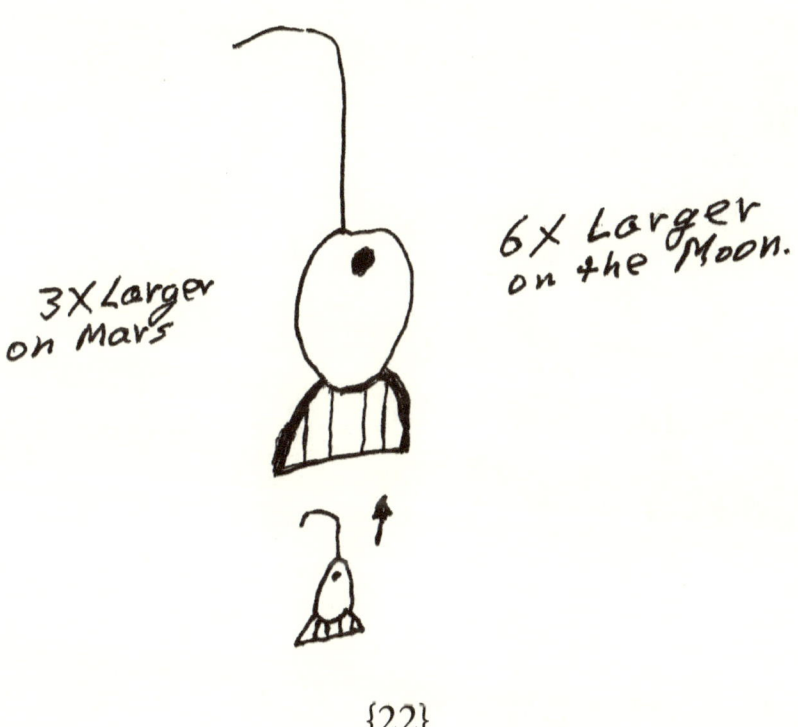

{22}

Fish Diagram

www.ingramcontent.com/pod-product-compliance
Lightning Source LLC
Chambersburg PA
CBHW021926170526
45157CB00005B/2197